AERO SERIES VOL. 20

GRUMMAN F8F BEARCAT

BY EDWARD T. MALONEY

Aero Publishers, Inc.
FALLBROOK, CALIFORNIA

We wish to express our appreciation and gratefully acknowledge the assistance of the following persons, who provided photographs.

U.S. Navy
Clay Jansson
Bill Fornoff
Frank Mormillo
Tom Piedimonte
Herb Nodine
Grumman Aircraft Corp.
Air Museum Archives
Merle Olmsted
Peter M. Bowers
Warren Shipp
William Swisher
William T. Larkins
Harold Martin
Jim Sullivan
Don Walsh

Library of Congress Catalog Card Number 75-96071

Printed and Published in the United States of America by Aero Publishers, Inc.

GRUMMAN F8F BEARCAT

By Edward T. Maloney

One of the most popular Navy fighters of all time is the Grumman "Bearcat." Born of battle. This tried and tested carrier fighter was the pride of the fleet from 1945 until it was retired with the coming of jet fighter aircraft.

The F8F was born in wartime secrecy. It was designed as a "Hellcat" fighter replacement and to out-perform the premier Japanese fighter the A6M-5 "Zeke" Zero Fighter.

In the summer of 1943 F6F-3 "Hellcats" were being turned out on Grumman production lines. Chief Grumman Engineer, William Schwendler, and his staff worked on Grumman design number 58. This new carrier fighter was designed to fill a major gap in current Navy fighters. This was to be a light weight interceptor in the low to medium altitude ranges, but with maximum maneuverability and speed. It was to be the approximate size of the pre-war F4F "Wildcat," but capable of hitting enemy aircraft in its climb up from carrier decks. It was a typical Grumman design — short, and stubby and powerful.

On November 27th, 1943, a contract was awarded Grumman Aircraft and Engineering Corp. for two XF8F-1 prototypes. The first XF8F-1, Bureau Number 90460, made its maiden flight June 25, 1944. It featured the first bubble canopy designed for a Navy fighter. This feature greatly improved visibility. The Bearcat utilized the tried and proven Pratt and Whitney R-2800 series engine with water injection.

On August 31st, 1944, the XF8F-1 was flown by Grumman test pilot, Robert Hall. He was very enthusiastic about the Bearcat's splendid performance. He stated he had never flown any aircraft quite like the Bearcat.

The first prototype crashed during initial testing and further tests were conducted with the number two XF8F-1.

A number of modifications were made to improve performance. A dorsal fin was added to the vertical fin to improve direction stability. The tail surfaces were rounded and the stabilizer was lengthened. The Bearcat sat in a rather nose-high attitude because it employed a 12 ft. 4 in. four-bladed propeller.

On October 6th, 1944, the first F8F-1 production order was placed. The Navy began to train pilots for the F8F program. By February 1945, the first twenty-three F8F-1 aircraft had been delivered. The Navy felt the Bearcat such an impressive fighter that a contract was made for 2,135 F8F-1's. The mass production capability of General Motors was recognized and its Eastern Aircraft Division was ordered to produce 1,876 F3M-1 Bearcats. The sudden end of the Pacific War prevented deliveries and none were built by Eastern Aircraft.

In October 1944, the XF8F-1 was delivered to the Navy's test center at Patuxent River, Maryland. It was located only a short distance from Washington, D.C. At this time the Joint Fighter Conference was taking place. Here, for the first time, were representatives from all the large American aircraft companies: Vought, Grumman, Northrop, Goodyear, Douglas, Curtiss, and United Aircraft Corp.

The Navy's newest, and current carrier fighters were on the flight ramps ready for combat trials. Represented were "Wildcats," "Corsairs," "Hellcats," the "Tigercat," and the new "Bearcat".

A special invited guest was a Mitsubishi A6M-5 "Zeke 52" Zero Fighter. This aircraft had just been captured intact on Saipan in June. It was brought here to be tested against the best American fighters. This Zero aircraft is now on display in The Air Museum at Ontario, California.

Charles Lindbergh, who was just back from flying a combat tour in the South Pacific, was a special invited guest pilot — as were members of the Army Air Force, and Royal Air Force.

Purpose of the conference was to pit the best American and Japanese fighter aircraft against each other in mock combat trials.

The prototype XF8F-1 was made available for flying trials by military and aircraft company test pilots. The results of the conference gave the XF8F-1 the highest percentage of pilots' votes for the best all-around fighter below 25,000 feet.

In November 1944, the second XF8F-1 prototype was completed. It was test flown by Grumman test pilot, Pat Gallo, in December. A small dorsal fin had been added to improve directional control. The prototype XF8F-1 models had Hamilton Standard Propellers, however the production Bearcats utilized Aero Production Propellers.

A production contract was soon signed for F8F-1's and a monthly production rate of 100 aircraft per month was set for the middle of 1945. As "bugs" were ironed out of the Bearcat design, the first XF8F-1 was sent to Langley Field for full-size wind tunnel tests.

Shipboard carrier trials commenced on February 17, 1945, aboard the USS Charger. The results were completely successful and the Bearcat was approved for carrier operation, including operation from the smallest escort carriers.

The first Navy Bearcat squadron, VF-19, was formed at Santa Rosa Naval Air Station, Santa Rosa, California, on May 21st, 1945. Commanding Officer was Lt. Cdr. Joseph Smith.

A training program was initiated as soon as the F8F-1's started to arrive from the Grumman factory. Carrier combat deployment was planned for late summer.

One of the chief problems encountered was with hydraulic system component fluid leaks and engine failures.

After these problems were overcome, VF-19 boarded the USS Takanis Bay, July 1945, for carrier qualifications. With these tests successfully completed, VF-19 deployed to the Pacific aboard the USS Langley on August 2, 1945. Commander Joe Smith's pilots of VF-19 were all eager to get into combat against the Zero, for they all believed they would become "Aces" on their first mission, however this was not to be. The B-29's of the 20th and 21st Bomber Commands had done their jobs well and Japan was reeling under heavy bombardment attacks.

Meanwhile, the Navy was forming the second Bearcat fighter squadron, VF-18, at North Island, San Diego, California.

In August, 1945, Japan surrendered and the Pacific War ended before the Bearcat could reach combat.

Production quotas were cut back. The F3M-1 contract was cancelled; however, the F8F continued to serve in the Navy's postwar program. A total of 770 aircraft were built on the first Navy-Grumman production contract.

The postwar Navy Air Force recovered as quickly as possible. The rapid changeover of wartime personnel to civilian life had its immediate effects on the Navy, but by 1946, four new Navy squadrons began converting to Bearcats. 1946 saw the introduction of the four 20mm cannon version, the F8F-1B model.

The F8F-1 had found a secure home in the postwar Navy plans. Nine fighter squadrons were in service with Bearcats. It would serve as the Navy's first-line carrier fighter until jet fighters were established on the Navy's aircraft carriers.

The next major Bearcat model, the F8F-2, began to come off the production line in early 1948. It utilized a 12 inch higher tail and a late model Pratt and Whitney R-2800-34 engine. Standard armament was four 20mm cannon. Production was planned for late 1947; however, it was early 1948 before this new model began to appear.

The F8F-2 had designed into it a unique feature called Automatic Engine Control (AEC). This unit regulated the throttle and the variable speed supercharger. With this system, manifold pressure regulation was accomplished entirely with one operation, which replaced both the conventional throttle control and supercharger control.

This time-saving device was to give Navy fighter pilots more free time for flying.

One unusual feature in the Bearcat design was the use of safety wing tips. The outboard section of the wings could be jettisoned, by the use of explosive bolts placed in the wing, if the aircraft wing became overstressed or encountered a high "G" load. This expendable wing section was located just outboard of the center aileron hinge. If one outer wing panel came off in a high speed dive, the pilot could blow off the opposite wing panel and still recover from his dive.

The F8F-2 had a revised engine cowl, plus internal changes. Sixty F8F-2P aircraft were produced for photographic purposes, two 20mm cannons were deleted to lighten the aircraft and improve its performance. The F8F-2P was intended for use as a long range land or carrier based photographic reconnaissance aircraft. It could carry a large number of cameras of the K-17 and K-18 type for vertical and oblique photography. The camera equipment was installed in the fuselage aft of the pilot's cockpit. Twelve F8F-2N night fighters were outfitted with radar. These would be ready for night carrier defense, should they be needed.

During the 1946 National Air Races at Cleveland, the Bearcat made a lasting impression on the attending crowds. Its spectacular rate of climb was beyond belief. The power vs. weight ratio was the best of any fighter to date, and the F8F literally jumped into the air.

On November 20th, 1946, the Bearcat returned to Cleveland to set a new national time to climb record: Lt. Cdr. M. W. Davenport flew a standard F8F-1 carrier fighter to 10,000 feet in 94 seconds. He took off after a run of only 115 feet. This record stood for ten years until broken by a modern jet fighter. However its short take-off distance could not be equalled by jet aircraft.

A special civilian version of the Bearcat was built for Major Al Williams, for promotion of Gulf Oil Products. It was some 3,000 lbs. lighter than its military brothers, and its powerful engine could deliver 3,000 hp., for short periods with water injection. Al Williams claimed he could do 500 mph at 19,000 feet altitude. He easily reached 10,000 feet in under 98 seconds during a number of air demonstrations.

In 1947 the Navy's "Blue Angels" demonstration team selected the Bearcat to replace its ageing Hellcat aircraft for airshow demonstrations. They flew these a number of years until the Navy jet fighter appeared.

By early 1948, the Navy had some twenty-four fighter squadrons equipped with Bearcats.

The Marines also flew the stubby little Bearcat. Three Marine bases were equipped with F8F's, and these were aircraft assigned to training squadrons. Aircraft were assigned to Cherry Point with VMFT-10, at Quantico with Aircraft Engineering Squadron 12, and at El Toro with VMFT-20. Marine pilots really appreciated the Bearcat's fantastic fighter performance. Its maneuverability was second to none. Its rapid take-off and climb performance left its opponents in a cloud of dust.

Marine and Navy Pilots loved to bounce Air Force P-51D "Mustang" fighters. The ensuing melee generally proved the Bearcats' superiority, much to the displeasure of Air Force Pilots trying to compete.

With the advent of the McDonnell FD-1's (FH-1) "Phantoms," and North American FJ-1 "Furys," the Navy began to use the F8F-1 for other duties. In 1945-6, a number of F8F-1's were modified as "Drone Control" aircraft. These aircraft were assigned to bases employing pilotless aircraft and target Drone test centers. Bearcats were also assigned to the Training Command as advance fighter trainer replacements for elderly Hellcats. As time went on, and more and more jet fighters entered fleet service, the number of F8F squadrons decreased.

In May, 1949, the last F8F-2 left the production line at Grumman Aircraft Corp.

During the Korean War the Bearcat was not considered suitable for operations by the Navy. F9F "Panthers" and F2H "Banshee" jet aircraft were the Navy's first line carrier fighters. The "Corsair" was selected to be the chief ground support propeller-driven fighter aircraft and so the Bearcat took second place status, however the Bearcat's career was not to end here!

In 1954, the United States came to France's aid, when 100 F8F-1 and 1B's were transferred to France and sent to Indo-China. Ironically, they were used in a role the U.S. Navy considered unsuitable for its use in Korea. The French Expeditionary Force placed the Bearcats into two ground support squadrons. They played an important part in supporting French ground units fighting the Viet Minh. The Bearcat's ability for short field take-offs and landings were especially beneficial to front line French Forces.

After the fall of Dien Bien Phu, and the French withdrawal, the remaining Bearcats were given to the Vietnamese Air Force. These Bearcats served with the 514th South Vietnam Fighter Squadron.

Thailand was the next to receive Bearcats, when the Royal Thai Air Force was given twenty-nine F8F-1B's by the United States. These aircraft carried an armament of four 20mm cannons. These Bearcats continued to serve well into the 1960's. There are still a number of Bearcats in Thailand and South Vietnam. These aircraft served as memorials and gate guardians to famous military bases.

By late 1955, the last F8F-2's were phased out of U.S. Navy Reserve bases and placed in stand-by storage at Litchfield Park, Arizona, and North Island, San Diego, California. Fifty F8F-1's were set aside at North Island for possible retention and use by France. In 1957 these aircraft were released and surplus sales were initiated by the Navy during 1958-1959.

The majority of the Bearcats held by the Navy were sold for scrap, however a few more were sold to civilians and for museum use. At least one was flown illegally out of the country to Cuba. At the present time, there are thirteen F8F's. Not included in this total is the F8F-2P held by the Air Museum at Ontario, California.

Although the number of Bearcats is diminishing, they are not yet gone! One, especially modified for air racing by Darryl Greenamyer, has won the Reno Air Race unlimited division for the past four years. The first Reno Air Race held in 1964, was won by a stock Bearcat sponsored by the late Bill Stead.

The Bearcat was a powerful brute. It was the smallest fighter that could have been designed around the powerful Pratt and Whitney R-2800 Double Wasp engine. It gave the Navy the edge over existing Japanese fighter types. Although it never attained air-to-air combat, it provided the Navy's first postwar line of defense.

No longer do these warriors return to their carriers. The Bearcat has served its last official mission, but it will always live on in the minds of the few who flew it and maintained it. Bearcat pilots are a privileged set. They surely will not forget the rapid acceleration and climbing abilities of this remarkable carrier fighter.

Editor's Note: As this book goes to press, Darryl Greenamyer in his highly modified F8F-2 has set the new world's land plane speed record for piston-engined aircraft. Greenamyer's aircraft is featured in the color section.

Number One Prototype XF8F-1 "Bearcat" Bu. No. 90460 was the star attraction of the Navy's Joint Fighter Conference held at Patuxent River, during World War II.

In-flight view of second XF8F-1 Bu. No. 90462. Note that the dorsal fin had not been added at the time this picture was taken.

After the crash of the first Bearcat, the second Prototype XF8F-1 carried on flight test work.

The F8F-1 was approved for aircraft carrier operations on February 17th, 1944, when this F8F-1 completed successful trials aboard the USS Charger (CVE-30).

Underwing bomb and rocket racks are clearly visible in this view of F8F-1 Bu. No. 94893.

F8F-1 Bearcat could carry three 500 lb. bombs, four 5" rockets and four 50 cal. machine guns.

First in-flight view published of the F8F-1 was this Bearcat ST-47. The leading recognition fournals of the world covered this aircraft in great detail.

F8F-1 Bu. No. 90447 was one of twenty-three pre-production Bearcats. Note highly polished wing.

Standard armament of F8F-1 was four 50 cal. Browning machine guns. Note outer wing bomb and rocket racks.

F8F-1 Bureau No. 94776, shown during 1947 Navy maneuvers at Santa Rosa, California Naval Air Facility.

Bearcat just catching the arresting cable aboard the aircraft carrier USS Antietam. Note red and yellow stripping on rudder trim tab and lower tail panel.

F8F-1 at the Naval Air Test Center, Patuxent River, Maryland for special test.

F8F-1 of Reserve Squadron at Glenview Naval Air Station. Large "V" letter on rudder signifies Glenview. Orange band around insignia signifies U.S. Naval Reserve.

F8F-1 Bearcat on the line at Naval Air Facility, China Lake — Inyokern, California. Aircraft served at station testing air-to-air and air-to-ground rockets. Note the two 5 inch H.V.A.R. rockets under wing.

Glenview based F8F-1 Bearcat with outer wing panels folded, this feature permitted easy storage aboard aircraft carriers.

Beautiful Bearcat Formation on the prowl from Glenview Naval Air Station. These F8F-1's are on a training mission. Note the practice bomb rack installation and long range fuel tank.

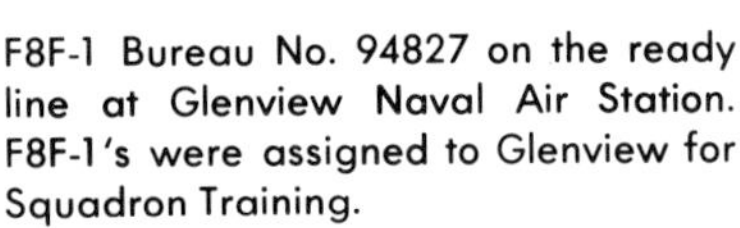

F8F-1 Bureau No. 94827 on the ready line at Glenview Naval Air Station. F8F-1's were assigned to Glenview for Squadron Training.

This F8F-1 is Naval Reserve aircraft from Olathe, Kansas. Note the roll-over structure under canopy for pilot protection and the 150 gallon external fuel tank.

Not all landings are happy landings. This Bearcat was attached to the Combat Information Center Pilots' School at Glenview Naval Air Station.

In a snow clad field! These F8F-1's are shown in front of base operations Glenview Naval Air Station near Chicago, Illinois.

Beautiful Bearcat formation is maintained by the Navy's precision aircraft demonstration team — the world-renown "Blue Angels".

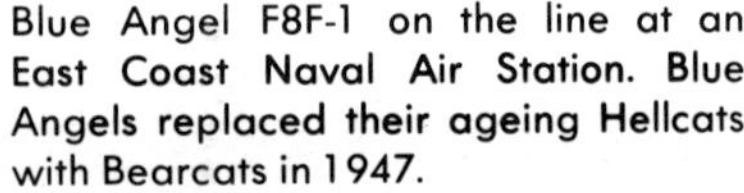

Blue Angel F8F-1 on the line at an East Coast Naval Air Station. Blue Angels replaced their ageing Hellcats with Bearcats in 1947.

Precision flying at its best became the key word of the Navy's "Blue Angel" flight demonstration team. The "Bearcat" permitted the "Angels" to keep their flying demonstrations in front of the grandstand at all times. A feat impossible to accomplish by later Blue Angel jet aircraft.

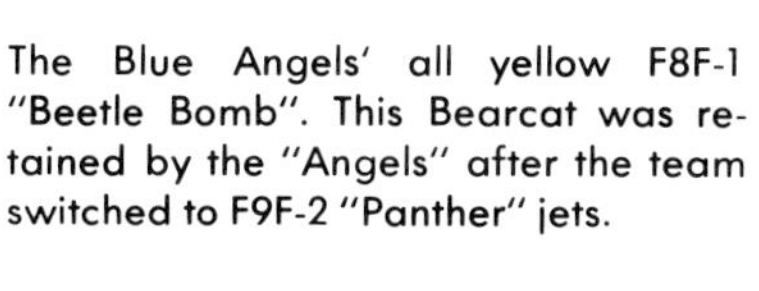
The Blue Angels' all yellow F8F-1 "Beetle Bomb". This Bearcat was retained by the "Angels" after the team switched to F9F-2 "Panther" jets.

Blue Angel F8F-1 "Beetle Bomb" with folded wings at the 1949 Cleveland National Air Races. Note the Navy AD-2 "Skyraiders" in the background.

Main instrument panel of the F8F-2 Bearcat Basic instrumentation was carried with flight instruments on the left and engine instruments on the right. Knobs to right and left of panel were gun changing controls.

Left-hand view of pilot's console. Throttle control is located at top left with rudder and elevator trim controls aft. Landing gear position indicator is just below throttle. Note fresh air control vent at extreme front.

Right-hand view of pilot's console. Volt meter is atop electrical switch panel with electrical distribution and radio control panels below. Sliding canopy control is located top center. Emergency hydraulic hand pump is by pilot's seat.

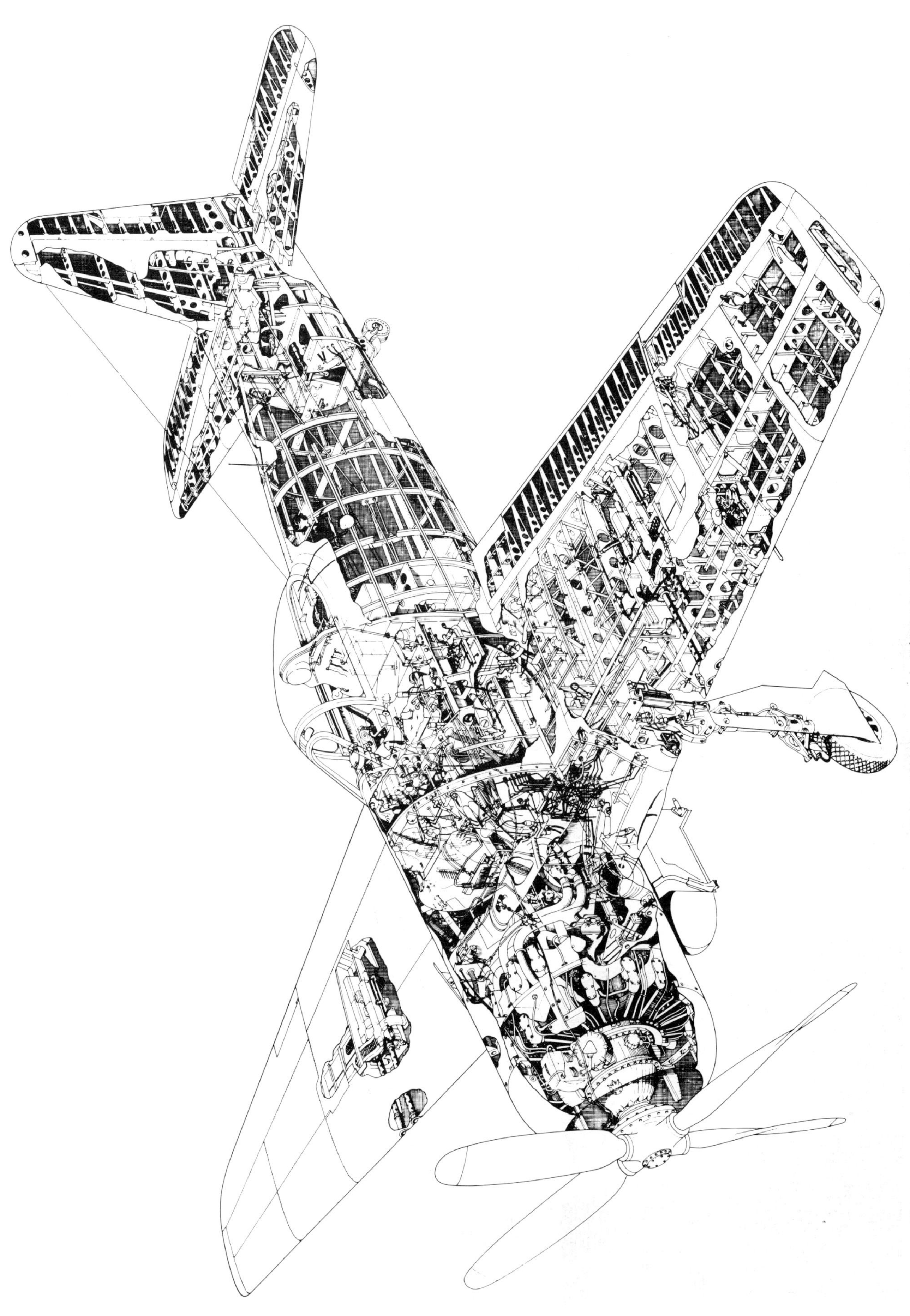

F8F-1 Bearcat with two 11.75 inch "Tiny Tim" air-to-ground rockets. "Tiny Tims" were the largest aircraft rockets ever fired from Naval fighter aircraft.

F8F-1 Bearcat Bureau No. 90438 with two 500 lb. bombs on wing racks. Note special experimental fins on 150 gallon external fuel tank converted to special Napalm bomb.

Grumman F8F-1 was a big attraction at the first postwar Los Angeles Examiner Air Show, 1946.

Number one Navy Blue Angel F8F-1 as flown by team leader Lt. Cdr. "Butch" Voris while leading the Navy Precision Aerobatic Team. The Bearcat became the replacement aircraft for the ageing F6F "Hellcat" previously used by the team. The F8F-1's of the Blue Angels made lasting impressions on Air Show crowds across the nation. The Bearcats were the last piston engine planes flown by the Blue Angels before converting to Navy F9F "Panther" jet aircraft.

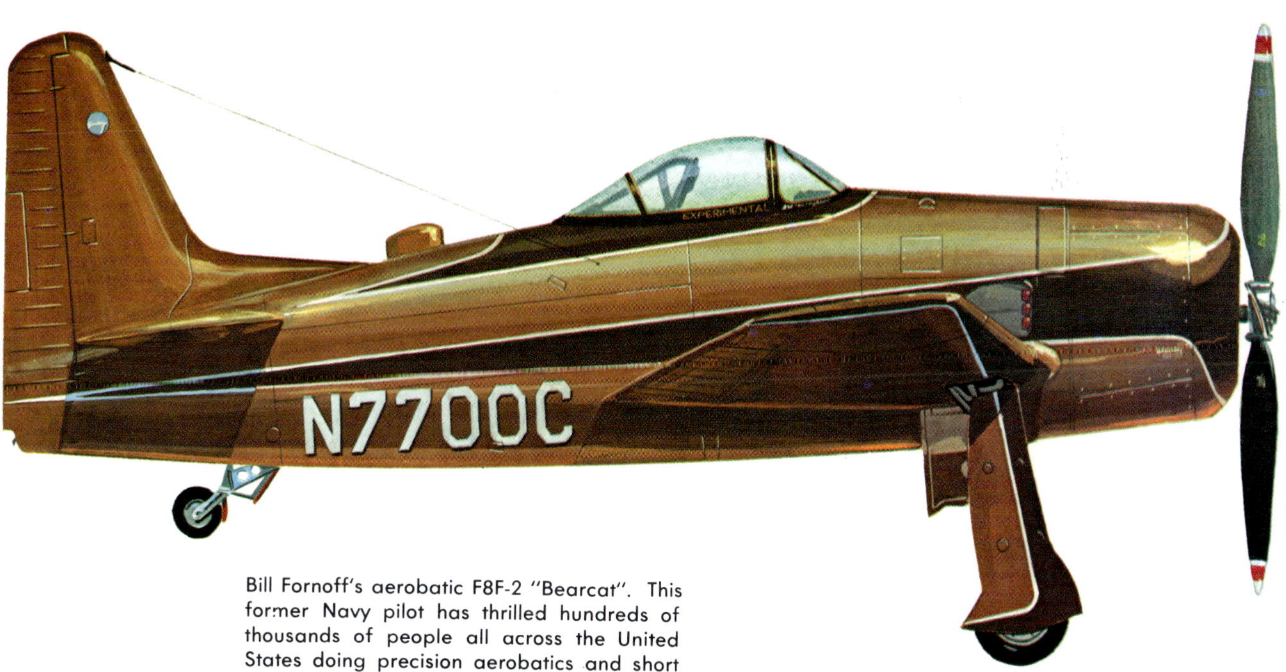

Bill Fornoff's aerobatic F8F-2 "Bearcat". This former Navy pilot has thrilled hundreds of thousands of people all across the United States doing precision aerobatics and short Navy carrier takeoffs and landings. His beautiful F8F-2 is painted a beautiful bronze-gold color with brown trim. A 10-coat wax finish is applied as the finishing touch.

James Dietz

Scale: 1:48

"THE RED RIPPERS"

Squadron Insignia have always played a very important part in Naval Aviation traditions. The Red Rippers' insignia was designed during the prewar years. This is one of the oldest and proudest of Naval Fighting Squadrons. This insignia design is on a shield of royal blue. The crest is the wild boar's head, which is claimed to have been taken from the label of a gin bottle. Directly under the boar's head are six gizmos of baloney. The two red ball circles signify bastardry and the red lightning bolt signifies speed. The meaning: — Gin Drunken, Baloney Slinging, Hot Rock Bas---ds!

Navy Grumman F8F-1 Bureau No. 94781 as flown by Ensign Walter Ohlrich while with Fighting Squadron VF-11, "The Red Rippers," flying off the aircraft carrier USS Ticonderoga.

Scale: 1:48

James Dietz

Winning Grumman F8F-2 "Bearcat" flown by Darryl Greenamyer at the 1967 Reno Air Race, Unlimited Division. He won 1st place prize money at a record speed of 396.2 m.p.h The aircraft qualified at 409.9 m.p.h. Greenamyer went on to recapture the world's land plane speed record (held by Germany since 1939) in this highly modified F8F-2 on Aug. 16, 1969, at Edwards AFB. His average speed was 480 mph, unofficially.

Grumman F8F-1 Bearcat of Royal Thai Air Force based at Don Maung Airport, Bangkok, Thailand. The Royal Thai Air Force received a number of these famous fighters in the late 1950's. The Squadron insignia is taken from the water buffalo — one of the world's most ferocious animals, to be found in Southeast Asia.

James Dietz

Scale: 1:48

Navy pilots give this F8F-1 Bearcat a thorough inspection.

Two damaged F8F-1's from USS Saipan. The one on left was damaged upon landing. Note bent propeller blades and missing tail wheel. Aircraft are from Carrier Air Group 5.

The Bearcat swung a 12′ 4″ Aero Products Propeller around a powerful Pratt and Whitney R-2800 "C" series engine of 2400 h.p.

This F8F-1 "Bearcat" served at China Lake Naval Ordnance Station, testing airborne rockets. Note 5 inch (H.V.A.R.) High Velocity Aircraft Rockets mounted under both wings.

The Navy F8F-IN Nightfighter Bureau No. 94819 contained a special wing rack mounted radar. Pilot had Radar Scope mounted in center of instrument panel.

F8F-2 Bu. No. 121726 above New York countryside on a factory test hop prior to delivery to Navy.

Short, stubby F8F-2 pulls up after completing a dive from altitude.

Surplus Navy F8F-2 "Bearcat" sits alongside F6F "Hellcat" awaiting disposition.

F8F-2 from the U.S. Navy Presidio at Monterey, California. Aircraft was stationed at Monterey to permit pilots to build up flying time.

This F8F-2 from the USS Leyte carried the standard armament of four rapid firing 20mm aircraft cannon. Note 100 gallon external fuel tank just between gear doors.

This head-on view of F8F-2 shows the Bearcat's extremely short wingspan of only 35 feet 6 inches.

F8F-2's began coming off the Grumman production lines in 1949. This model utilized the Pratt and Whitney R-2800-30 W engine.

F8F-2 Bu. No. 121714 at rest on the ramp at North Island Naval Air Station.

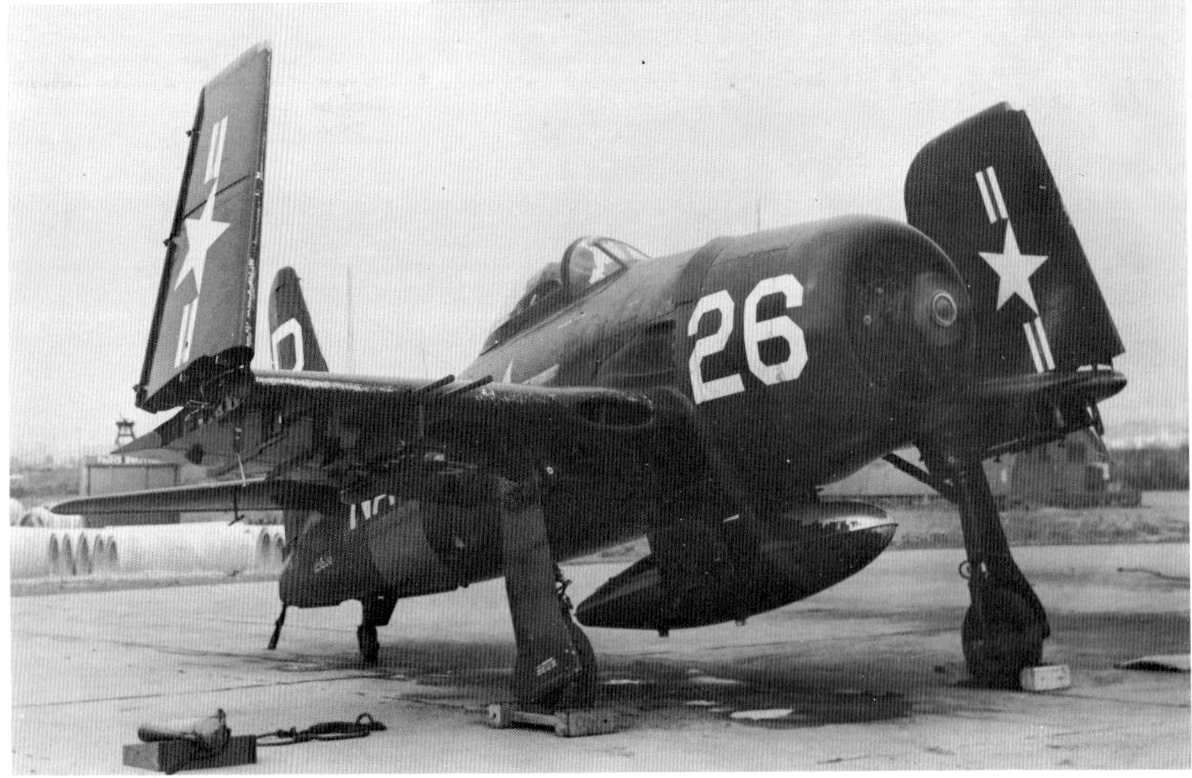

This F8F-2 was based with the Naval Reserve at Denver. Mechanic is running up engine in preparation for flight.

F8F-2 from Monterey Naval Air Facility on a cross country flight to Miramar Naval Air Station.

Striking view of F8F-2 from Naval Air Reserve Center, Denver, Colorado. The Bearcat was a challenge to fly and Reserve pilots rated it as our top postwar carrier fighter.

Surplus F8F-2 on the line awaiting disposition. Aircraft was originally part of the Naval Reserve Squadron at Norfolk, Virginia.

This view of F8F-2 clearly shows the external bomb and rocket racks under the Bearcat's wings.

This F8F-2 served with the Naval Air Reserve Squadron at Denver, Colorado. Navy assigned letter codes to aircraft to signify home stations. "P" for Denver, "V" for Glenview, "K" for Olathe, Kansas, etc.

Parade of Grumman aircraft at the Ontario Air Museum includes a host of Navy aircraft. Right to left "Avenger," "Hellcat," "Wildcat," "Bearcat," and Curtiss "Helldiver".

This Quantico based Marine F8F-2 is shown staging through Dallas, Texas on a cross country flight. The Marines had three stations of Bearcats, AES-12 Aircraft Engineering Squadron at Quantico, Va. VMFT-10 at Cherry Point, N.C., and VMFT-20 at El Toro, California.

Navy F8F-2N Night Fighter undergoing evaluation at Naval Air Test Center, Patuxent River, Maryland.

Flight deck load of F8F-1 Bearcats aboard USS Philippine Sea.

On the flight line at the Ontario Air Museum. F8F-2 Bearcats are made ready for flight. F8F-2 Civil Registration No. N1111L was later modified for air racing by Darryl Greenamyer.

Four F8F-1's on carrier flight deck beside the larger F6F-5 "hellcat".

F8F-1's taxi down carrier deck under direction of skillful handling of carrier deck hands in preparation for taking off.

Navy deck hand on wing guides pilot down long carrier deck. Note all wood top deck on carrier. Bearcat pilot awaits flag to commence takeoff.

Carrier Code letters "PS" were painted on tails of these F8F-1's in 1947, to signify USS Philippine Sea.

F8F-2 from Navy Utility Squadron 7 Miramar Naval Air Station.

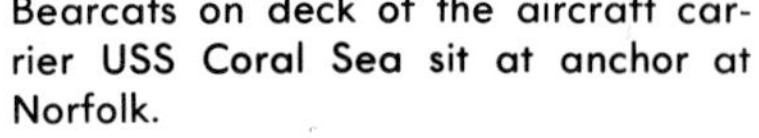

Bearcats on deck of the aircraft carrier USS Coral Sea sit at anchor at Norfolk.

F8F-2 Bearcat on the flight line at North Island Naval Air Station.

Bearcat employed use of safety wing tips which could be jettisoned by use of explosive bolts placed in the wing if the aircraft wing became overstressed or encountered high "G" loads.

Beautiful in-flight view by Wm. T. Larkins of Lt. Cdr. D. C. "Whiff" Caldwell's Bearcat fighter from the carrier USS Boxer.

Colorfully marked F8F-1 fighter of Cdr. H. E. Cook Jr., Commander of Air Group 19 from carrier USS Boxer.

Carrier deck view of four smart looking F8F-2's being prepared for takeoff during cruise aboard USS Midway.

Two rare Bearcats flying formation: F8F-2N in foreground and F8F-2P photo-recon aircraft in background. Note radio whip antennae.

Deadly air-to-air view of cannon armed F8F-1B in skies above Grumman Aircraft factory.

Beautiful carrier deck view of F8F-2 "Bearcat" ready for takeoff and awaiting Landing Signal Officer's flag aboard USS Midway.

Three Ontario Air Museum Naval fighters taxi out for flight to Navy airshow at Miramar NAS. Left to Right, "Bearcat" — "Hellcat" — "Corsair".

Bearcat flashes by camera at speed of better than 400 m.p.h.

Ontario Air Museum's Bearcat taxis out for flight to Navy Air Show at North Island NAS. Note UHF antenna aft of canopy.

Bearcat shows off its small angular lines in this flight view. It was short, stubby, and powerful.

Bearcat sits high on the ground. Pilot must make "S" turns to see while taxiing.

Majority of remaining Bearcats in the United States are the late F8F-2 model.

The Bearcat was the first Navy Fighter to have an all bubble canopy with 360 degree visibility.

The Bearcat succeeded the Hellcat (with folded wings) as main fleet carrier fighter following World War II. By 1949 some twenty-four Navy fighter squadrons were equipped with Bearcat fighters.

Running up F8F-2 Bearcat in preparation for taking off. This model featured the powerful 2400 h.p. Pratt and Whitney R-2800-30W engine with water injection.

Check pilot A. L. Redick takes off in Bearcat, heading for Navy Air Show. Note landing gear retract sequence: left landing gear retracts ahead of right gear.

Like a bullet shot from a gun! The Bearcat leaps into the air after a short takeoff run.

Stubby F8F-2 was brute of power. The biggest engine available was mounted on the smallest fuselage design possible.

Bearcat lifts off runway after short takeoff run. F8F had good controllability throughout various speed ranges.

End of the line! Group of F8F'2's awaiting scrap metal smelter at North Island NAS, San Diego.

Major Al Williams flying his Grumman G-58A "Gulfhawk 4," Aircraft was used to promote Gulf Oil Products.

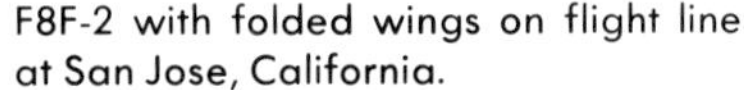

F8F-2 with folded wings on flight line at San Jose, California.

Bill Fornoff performs one of the most spectacular flight demonstrations that can be viewed at airshows today. He is shown here doing a wing over.

Bill Fornoff taking off in his beautiful F8F-2 Bearcat, for Naval Air Show.

Civilian registered Bearcat N7700C at Torrance, California.

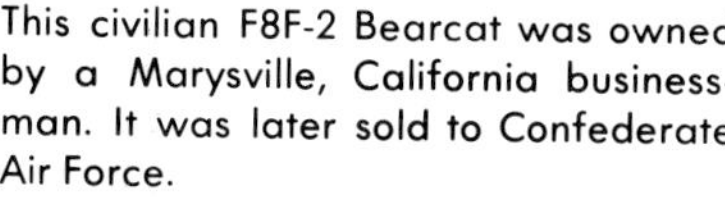

This civilian F8F-2 Bearcat was owned by a Marysville, California businessman. It was later sold to Confederate Air Force.

Bill Fornoff flying inverted in his gold-bronze painted F8F-2. His aerial flight demonstration in the Bearcat is a real crowd pleaser.

One of the three remaining F8F-1 model "Bearcats" in the United States. This one is currently owned by an air-line pilot in Illinois.

This F8F-1 was owned by Grumman Air-craft Corp. and was used for special projects and research work. Note spe-cial radio antenna.

F8F-1 N 700A was acquired by Cornel University and used by their flight test laboratory testing various electronic devices.

C. F. Christopher's Grumman F8F-2. This New York based Bearcat is still flying to various East Coast Air Shows.

Beautiful all yellow and black trimmed Bearcat owned by Stan Krazet.

This F8F-2 has a P-51H aeroproducts propeller.

Stan Krazet's F8F-2 at the 1966 Reno Air Races.

Remains of Grumman F8F-2. Pilot error claimed this once beautiful F8F-2 Civil register number N1031B at Valparaiso, Indiana.

Bearcat racer once owned by John Church and flown by Chuck Lyford for air show demonstration flights.

Chuck Lyford coming over field at better than 400 m.p.h. Small and powerful F8F enables pilots to keep air show demonstration in view of the spectators at all times.

The late Robert Kucera's F8F-2 Bearcat. Kucera used this aircraft in his high altitude photographic survey work.

Kucera had a special side door built into his Bearcat to permit two men to operate aerial cameras for his flight survey work. Note special structure built around door frame.

Ground view of Bill Fornoff's F8F-2 Bearcat at the Lancaster Air Races.

The Fornoff F8F-2 is always kept in tip-top shape. It has appeared throughout the United States at most major air shows. Note non-standard pneumatic tail wheel.

Beautiful Gold-Bronze color scheme on Bill Fornoff's aerobatic Bearcat can be seen at a great distance.

With Mira Slovak as pilot this Smirnoff sponsored Bearcat won the 1964 Unlimited Division, Reno Air Race.

Race pilot Mira Slovak rounds pylon during Reno Air Races.

F8F-2 "Tom Cat" as entered in the 1964-65-66 Reno Air Races. Flown by Navy Commander Walter Ohlrich, Jr.

Air Race pilot Walter Ohlrich taxis Bearcat to take off in the 1968 Unlimited Division Reno Air Race. This is the only race in the United States where the public may see the big bore "Bearcats" perform.

A broken ignition harness prevented Commander Ohlrich from competing in the 1968 Reno Trans-Continental Air Race. Note the addition of two special 200 gallon external fuel tanks.

F8F-2 Bearcat as modified by Darryl Greenamyer for Reno Air Races. Darryl removed some 1800 lbs. to lighten ship for air racing.

Darryl Greenamyer's 1967 Reno Air Race winner. Note special Smirnoff color scheme: all white with blue metallic trim.

Wings were cut down on Greenamyer Bearcat to reduce drag for air racing. This view shows special AD-4 "Skyraider" propeller which was added.

Bob Kucera flew this Bearcat to 1st place in the 1968 Reno Consolation Air Race.

John Church flew this F8F-2 Bearcat in the 1966 Reno Air Races.

F8F-1 Bearcat of Royal Thai Air Force at Don Maung Airport, Bangkok, Thailand.

Royal Thai Air Force F8F-1 was one of 29 supplied by the United States to assist Thailand's Air Force.

One of the few remaining F8F-1B "Bearcats" Bu. No. 121510 of the South Vietnamese Air Force. This one is on display outside Bien Hoa military training station.

Sixty F8F-2P Photo-Recon carrier aircraft were built. Note camera window just aft of wing.

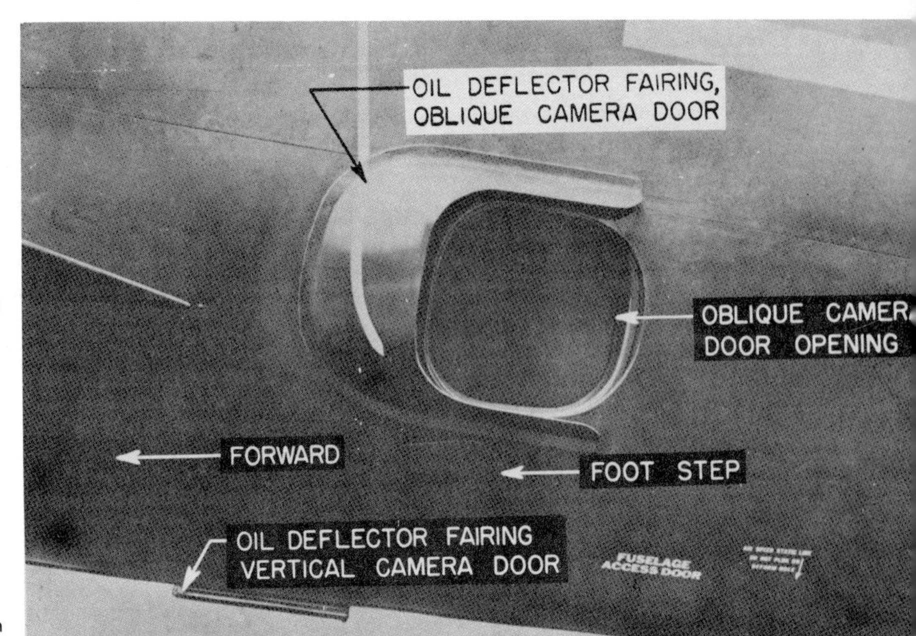

Close-up detail of F8F-2P oblique camera window.

Beautiful in-flight view of F8F-1 from USS Saipan (CVL-48).

BEARCAT TECHNICAL DATA

QUANTITY MFG.	MODEL	SERIAL NUMBER	MISSION
2	XF8F-1	90460 & 90462	Prototype
23	F8F-1	90437 & 90459	Pre-Production
746	F8F-1	94752 - 95497	Production Model
226	F8F-1B	Intermittent Numbers 94972 - 95498 121463 - 121522 122087 - 122152	Heavy Armament Fighter
(1)	F8F-1C	90440	Heavy Armament Fighter
(Several)	F8F-1D		Modified F8F-1 to Drone Control Aircraft
(2)	XF8F-1N	94812 & 94819	Night Fighter Prototype
(13)	F8F-1N	95034 95150 95182 95206 95230 95129 95161 95191 95214 95140 95171 95198 95222	Night Fighter
(2)	XF8F-2	95049 & 95330	Prototype
365	F8F-2	121523 - 121792 122614 - 122708	Carrier Fighter
(Few)	F8F-2D		F8F-2 to Drone Control Aircraft
12	F8F-2N	121549 - 121550 121575 - 121579 121601 - 121605	Night Fighter
60	F8F-2P	121580 - 121585 121606 - 121611 121632 - 121637 121658 - 121663 121684 - 121689 121709 - 121714 121734 - 121739 121758 - 121763 121770 - 121775 121785 - 121790	Photo-Recon Aircraft
1	G-58	N700A	Grumman Aircraft Corp. Research and Test Aircraft. Built from Spare Parts.
1	G-58A	NL-3025	Al Williams' Famous "Gulfhawk 4," High Performance and Demonstration Airplane.

1436 Total Mfg.

BEARCAT SPECIFICATIONS

MODEL	SPAN	LENGTH	HEIGHT	ENGINE	HORSE-POWER	MAX. SPEED	CEILING	ARMAMENT	NORMAL GROSS WEIGHT
XF8F-1	35′6″ (23′3″)	27′8″	13′2″	R-2800-22w	2100 hp.	424 mph. @ 17,300 ft.	33,700 ft.	(4) 50 cal. (2) 1000 Bombs & (4) 5″ Rockets	8,800 lbs.
F8F-1	35′6″ (23′3″)	27′8″	13′2″	R-2800-34w -32w	2100 hp.	421 mph @ 19,700 ft.	38,900 ft.	(4) 50 cal. (2) 1000 Bombs & (4) 5″ Rockets	9,600 lbs.
F8F-1B	35′6″ (23′3″)	27′8″	13′2″	R-2800-34w	2100 hp.		38,900 ft.	(4) 20mm cannons (2) 1000 Bombs & (4) 5″ Rockets	10,000 lbs.
F8F-1N	35′6″ (23′3″)	27′8″	13′2″	R-2800-34w	2100 hp.		38,800 ft.	(4) 50 cal. (2) 1000 Bombs & (4) 5″ Rockets	9,900 lbs.
XF8F-2	35′6″ (23′3″)	27′6″	13′8″	R-2800-30w	2400 hp.		40,000 ft.	(4) 20mm cannons (2) 1000 Bombs & (4) 5″ Rockets	9,800 lbs.
F8F-2	35′6″ (23′3″)	27′6″	13′8″	R-2800-30w	2400 hp.	447 mph @ 28,000 ft.	40,700 ft.	(4) 20mm cannons (2) 1000 Bombs & (4) 5″ Rockets	10,400 lbs.
F8F-2N	35′6″ (23′3″)	27′6″	13′8″	R-2800-30w	2400 hp.		40,500 ft.	(4) 20mm cannons (2) 1000 Bombs & (4) 5″ Rockets	10,600 lbs.
F8F-2P	35′6″ (23′3″)	27′6″	13′8″	R-2800-30w	2400 hp.	450 mph @ 28,000 ft.	41,000 ft.	(2) 20mm cannons (2) 1000 Bombs & (4) 5″ Rockets	10,100 lbs.